4th Grade Math
Volume 4

ISBN: 978-1-939796-85-1

Table of Contents

Fractions and Mixed Numbers

Key Vocabulary

mixed number

improper fraction

How many Pizzas?

Match the mixed numbers with the pizzas.

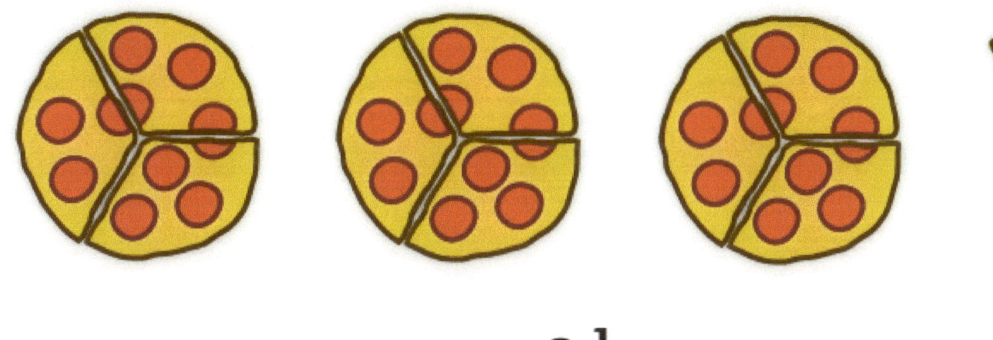

$$3\frac{1}{3}$$

A whole number and a fraction is called *a mixed number*.

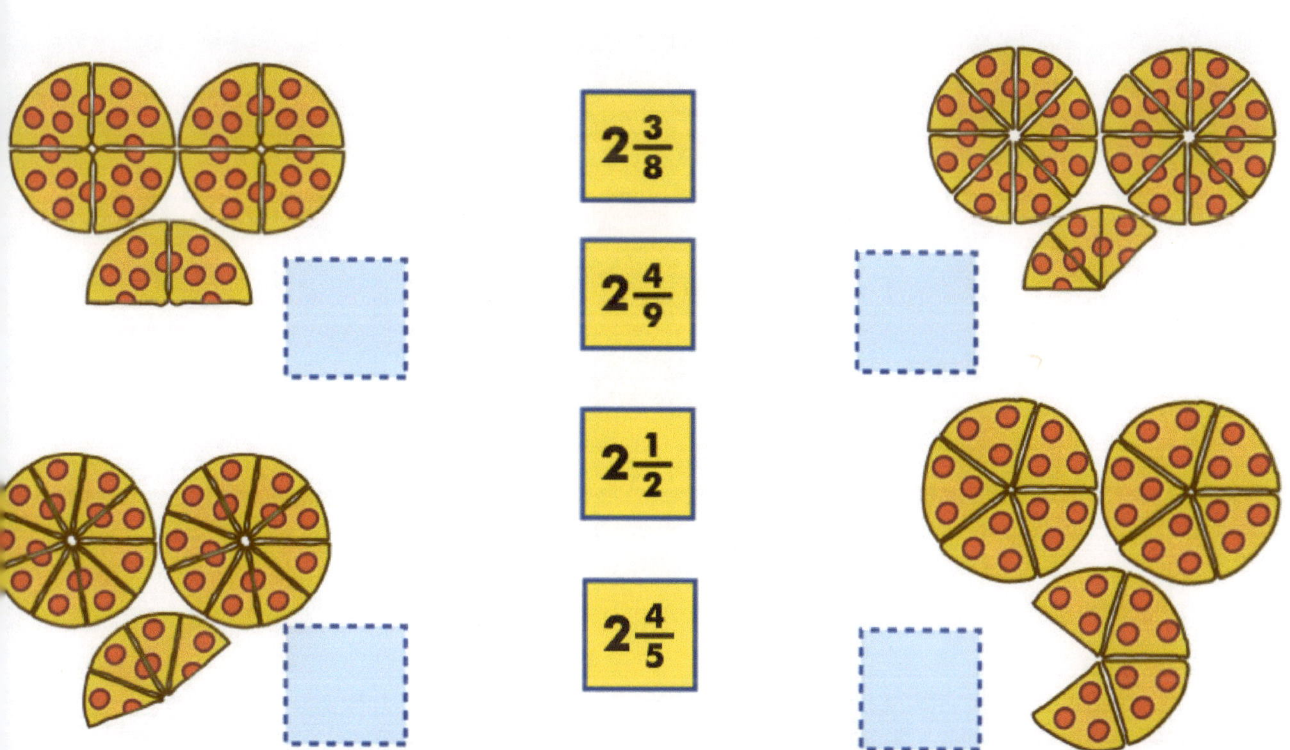

An *improper fraction* has a numerator equal to, or larger than, its denominator.

How many thirds in this improper fraction?

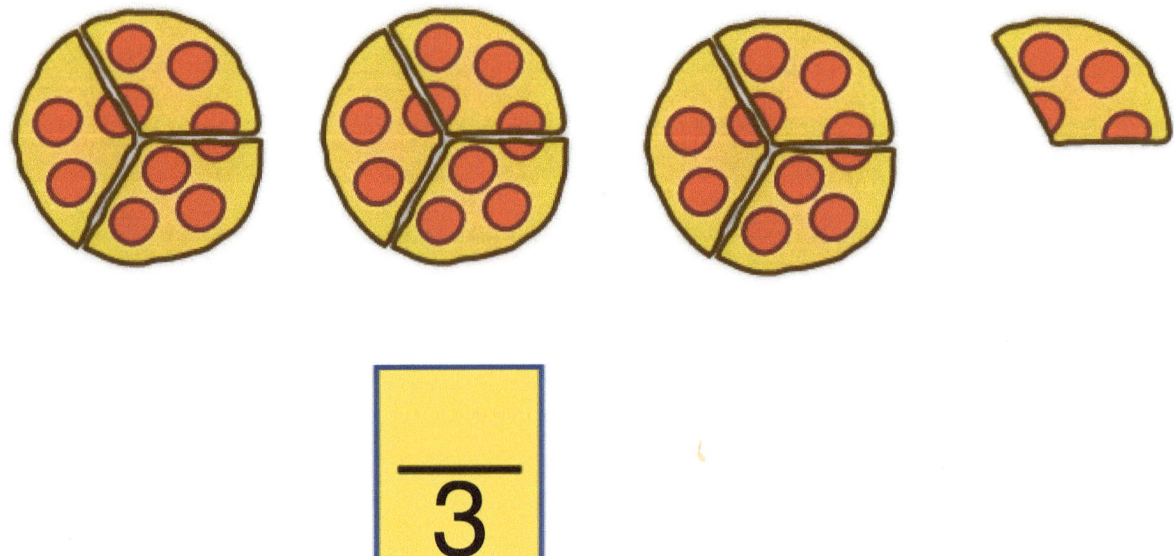

$$\frac{}{3}$$

Match the improper fractions with the pizzas.

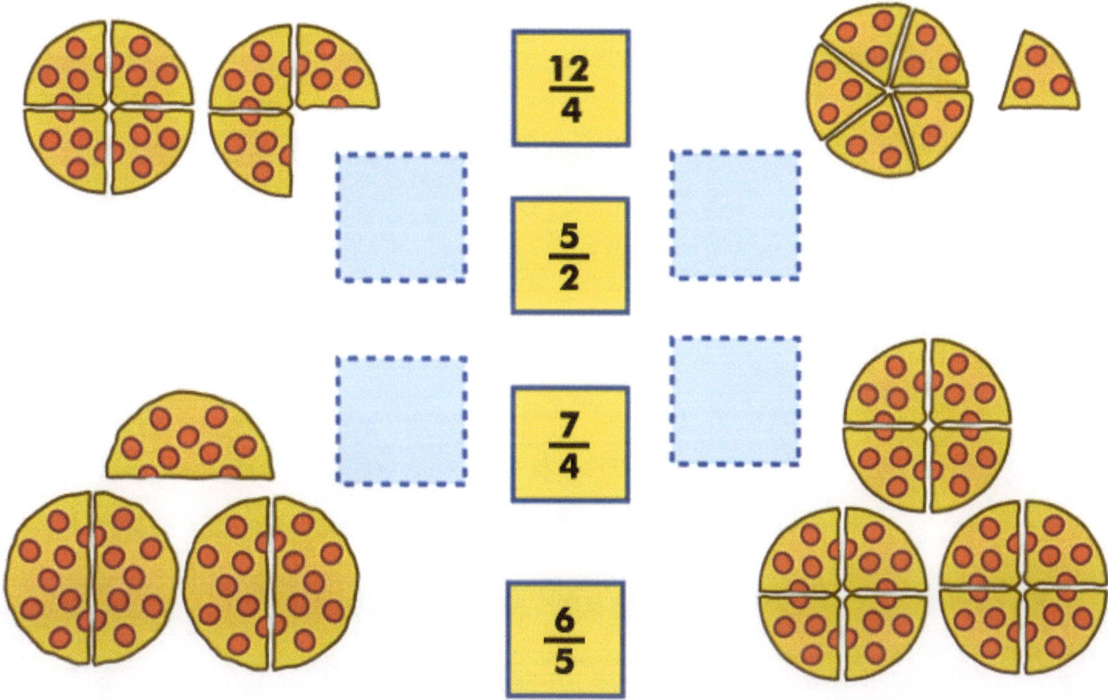

There are two methods for converting a mixed number to an improper fraction.

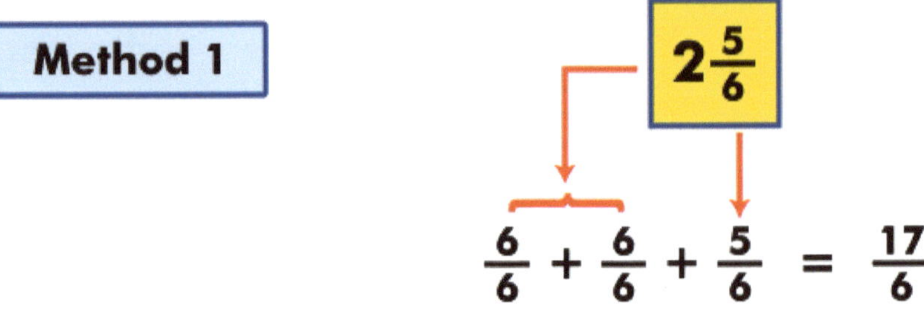

There are two methods for converting and improper fraction to a mixed number.

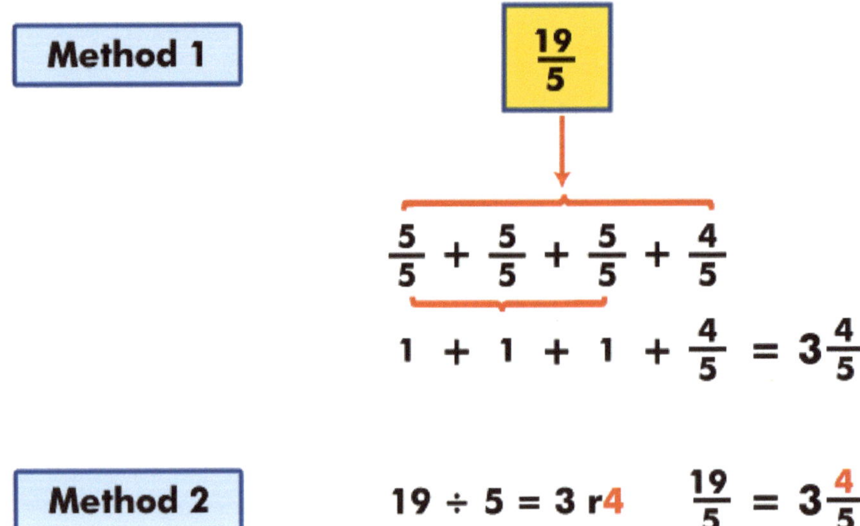

Convert these mixed numbers to improper fractions and these improper fractions to mixed numbers.

$$3\frac{4}{5} = \underline{\quad}$$

$$3\frac{2}{7} = \underline{\quad}$$

$$\frac{37}{5} = \underline{\quad}$$

$$\frac{56}{42} = \underline{\quad}$$

Write a number sentence for our Super Bowl party.

We bought this much pizza.   We had this much left over.

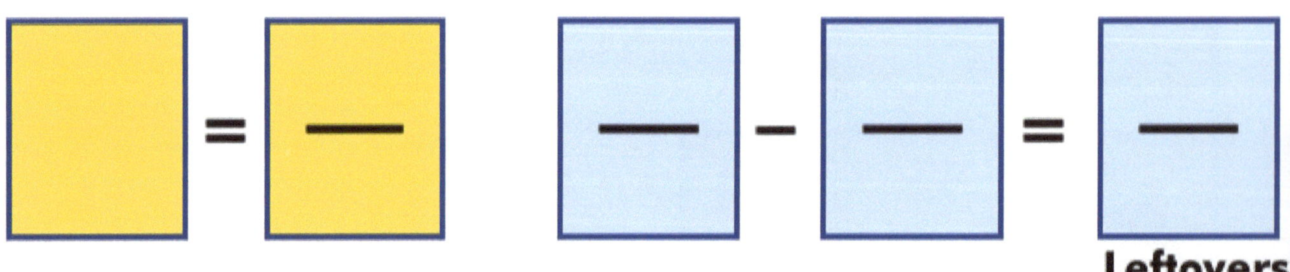

Leftovers

Fractions and Mixed Numbers Quiz

1   **True or false?** $3\frac{2}{3} = \frac{17}{5}$

2   **What is the value of x?**

- **A** $\frac{3}{4}$
- **B** $\frac{12}{4}$
- **C** $\frac{15}{4}$
- **D** $\frac{7}{2}$

3   **What is the value of y?**

- **A** $\frac{40}{8}$
- **B** $\frac{41}{8}$
- **C** $\frac{20}{4}$
- **D** $\frac{19}{4}$

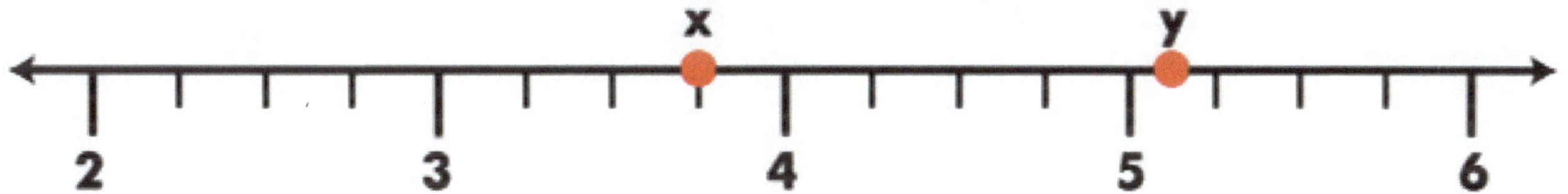

4 $\frac{47}{9} = 5\frac{?}{?}$

Add and Subtract Fractions

Key Vocabulary

fraction

numerator

denominator

equivalent fractions

Tangram

Equivalent Fractions

Fill in the blanks.

Can you fit these pieces in the whole and then complete the equation.

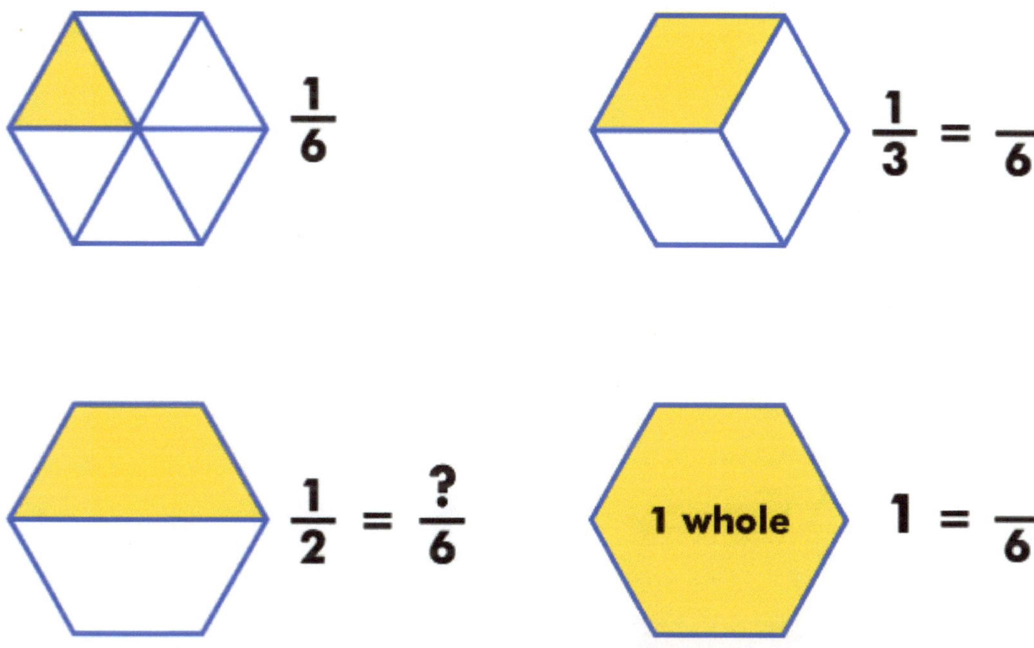

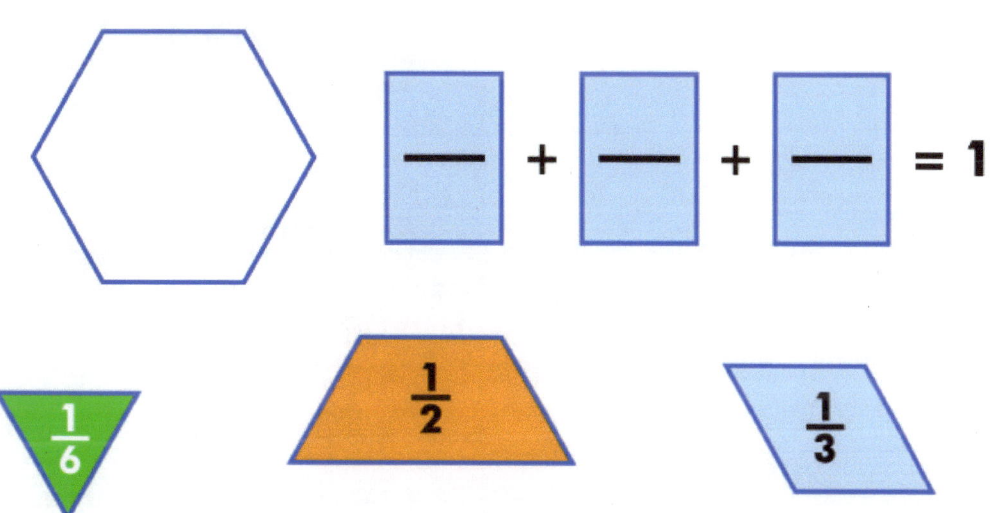

Find combinations that sum to one half.

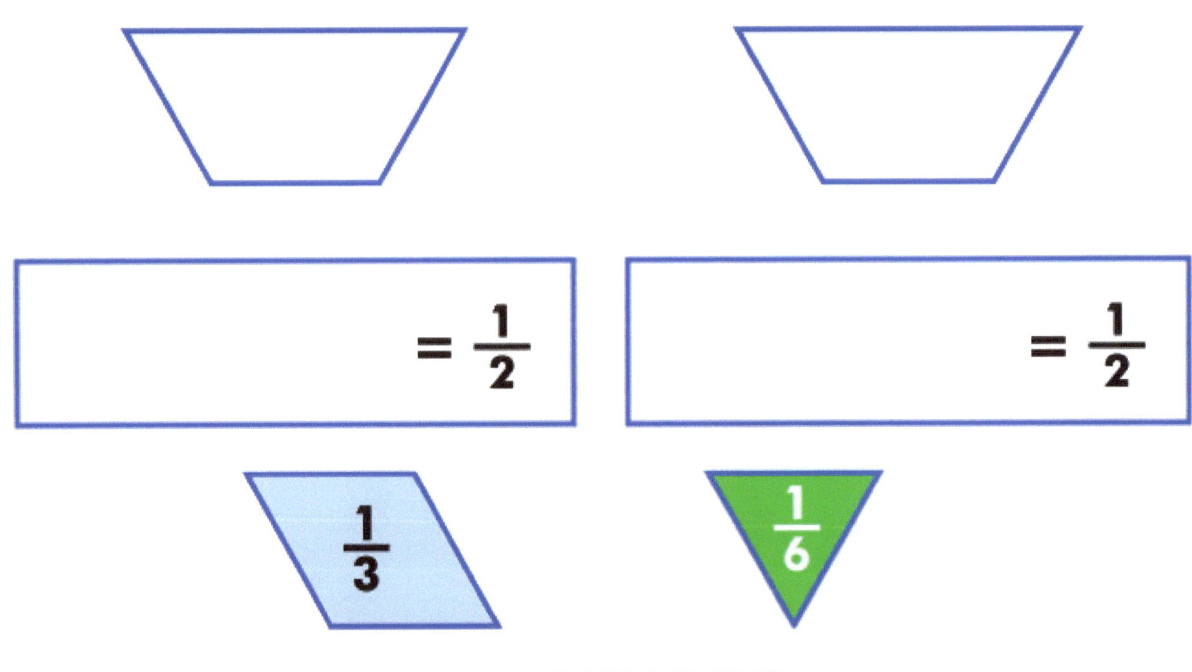

Identify each of the fraction pieces.

One has been completed to get you started.

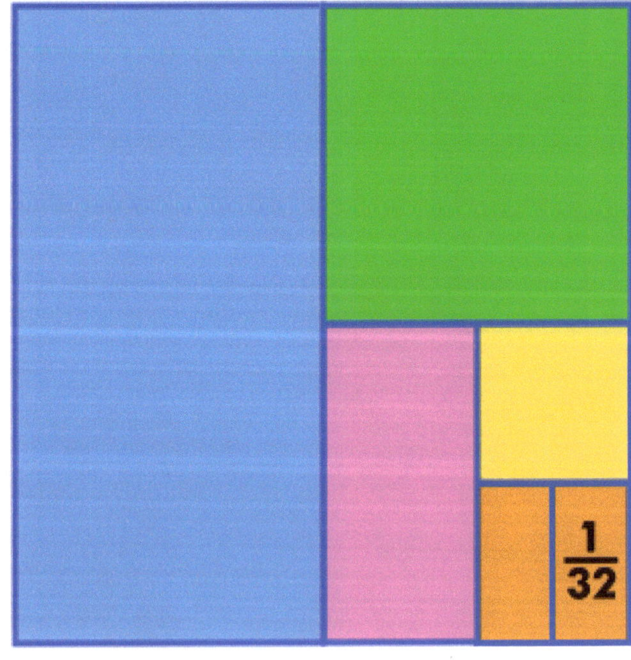

Model solutions to these fraction problems.

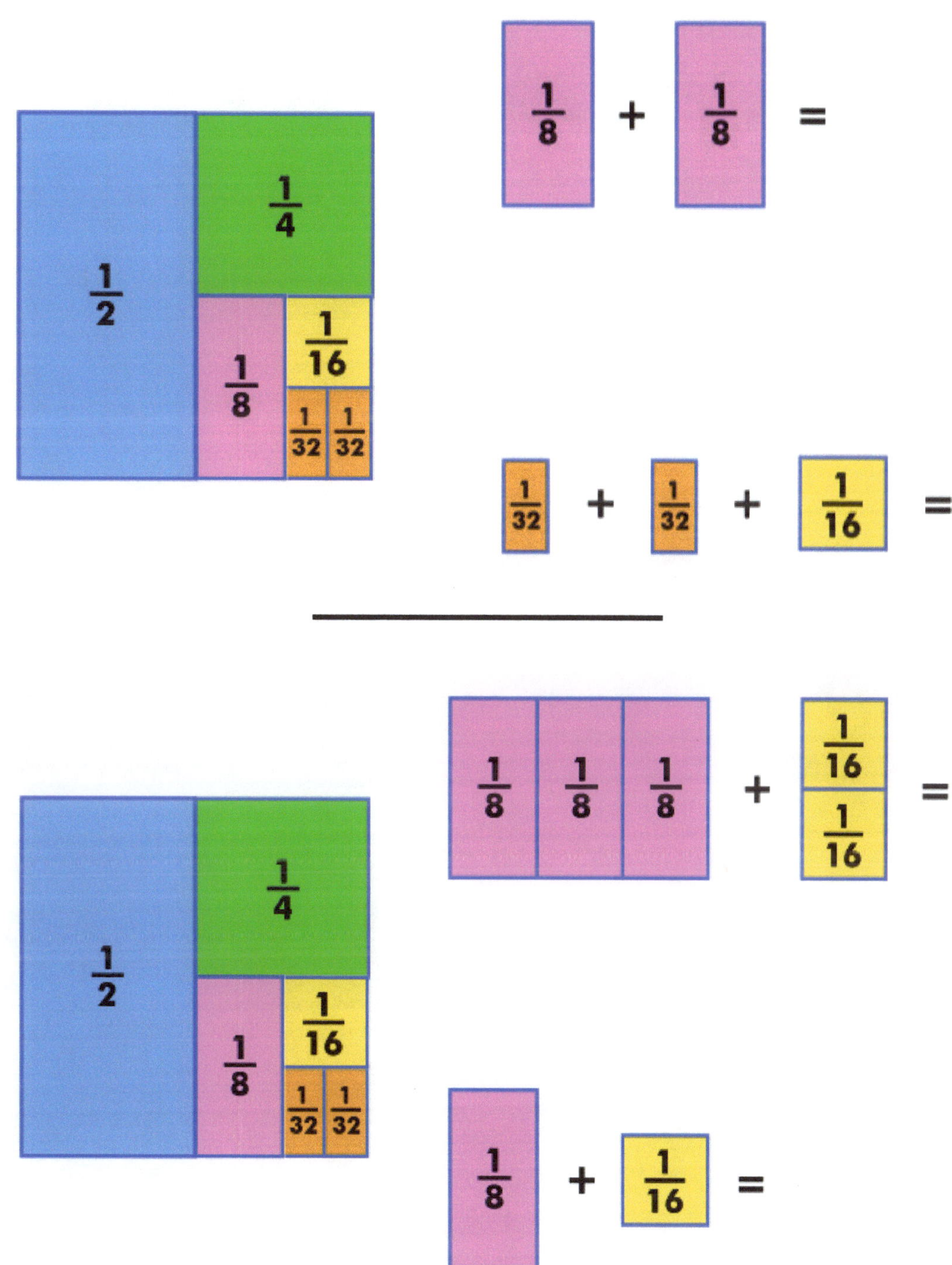

Add the fractions.

$$\frac{1}{2} \quad + \quad \frac{3}{8} \quad = \quad \boxed{?}$$

$$\frac{1}{4} \quad + \quad \frac{3}{8} \quad = \quad \boxed{?}$$

$$\frac{2}{16} \quad + \quad \frac{3}{8} \quad + \quad \frac{1}{4} \quad = \quad \boxed{?}$$

Add and Subtract Fractions Quiz

1   **True or false?** $\dfrac{1}{2} = \dfrac{2}{4}$

2 $\dfrac{1}{2} + \dfrac{1}{8} = ?$

- **A** $\dfrac{3}{4}$
- **B** $\dfrac{1}{4}$
- **C** $\dfrac{3}{8}$
- **D** $\dfrac{5}{8}$

3 $\dfrac{1}{4} + \dfrac{3}{8} = ?$

3 $\dfrac{1}{4} + \dfrac{2}{16} + \dfrac{1}{8} + \dfrac{7}{16} = ?$

Patterns and Functions

Key Vocabulary

pattern

rule

function

table

expression

operation

Continue the pattern and find the rule of the pattern.

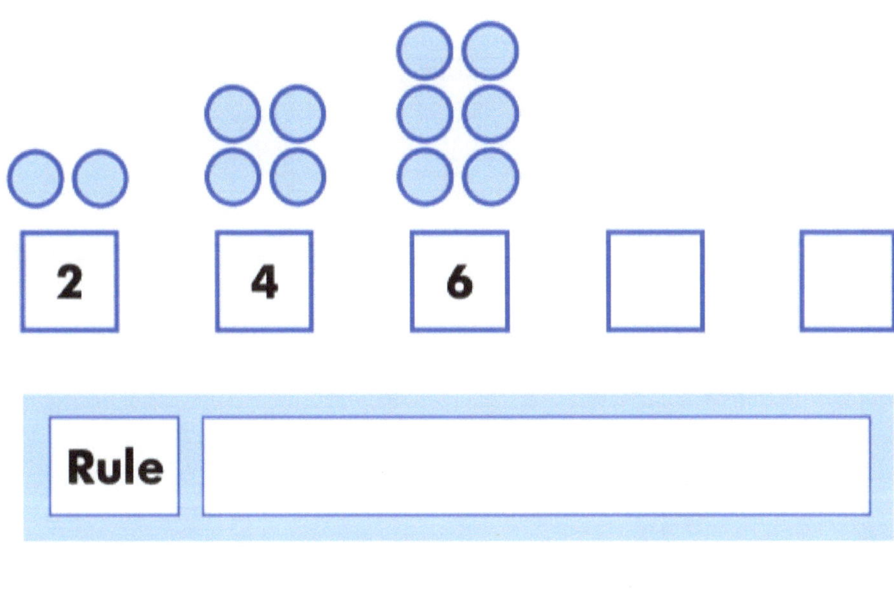

Rule	

____

Growing Patterns

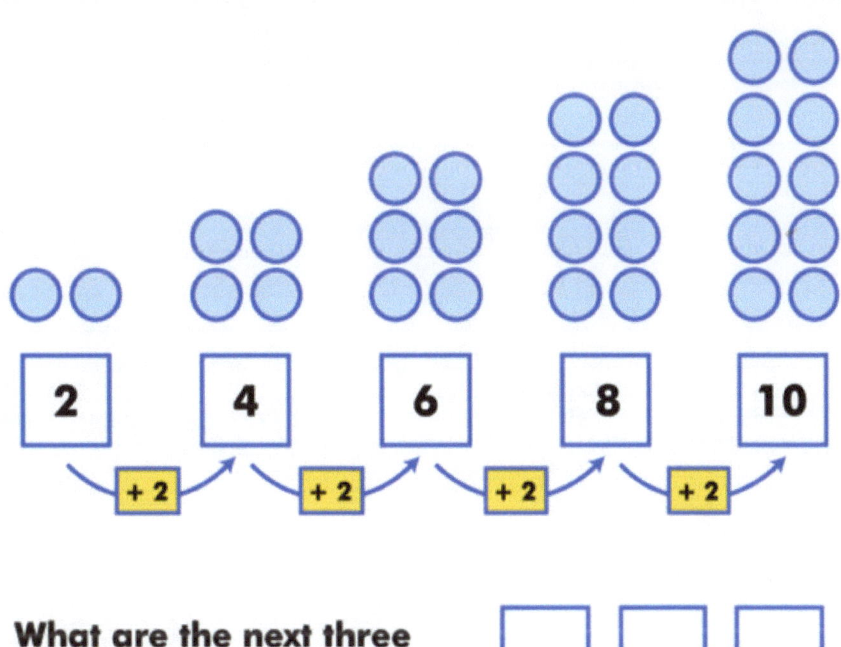

What are the next three numbers in the pattern?

Continue the pattern to complete figure 4.

What is the rule for the patterns above? _______________________

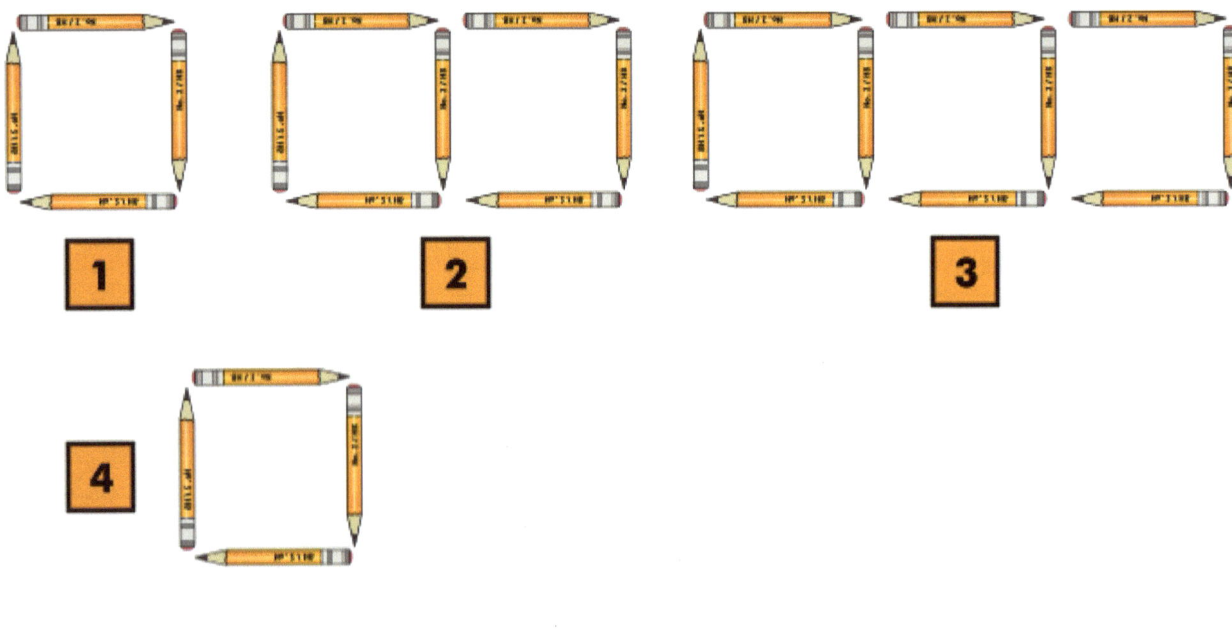

____

Complete the pattern and find the rule.

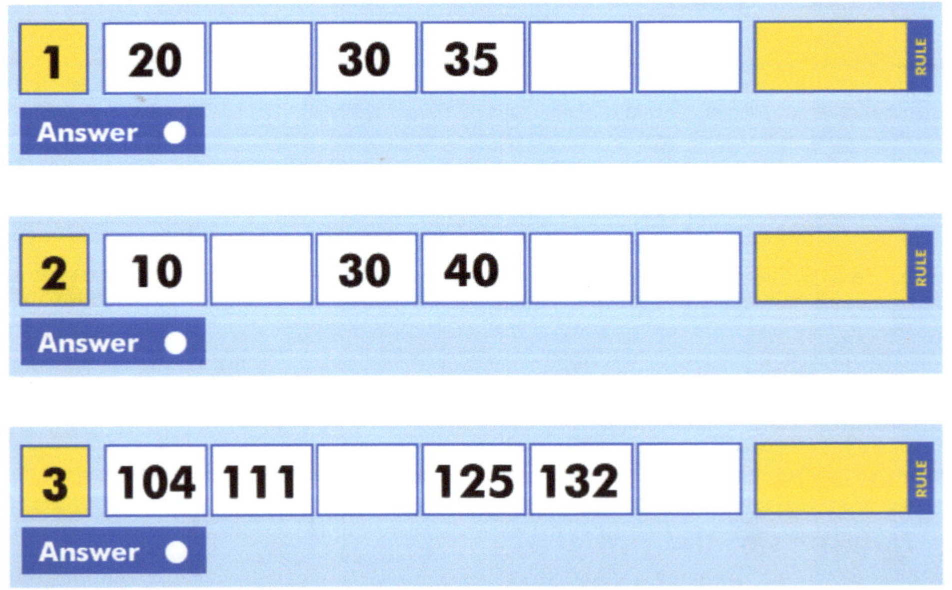

Complete the pattern and find the rule.

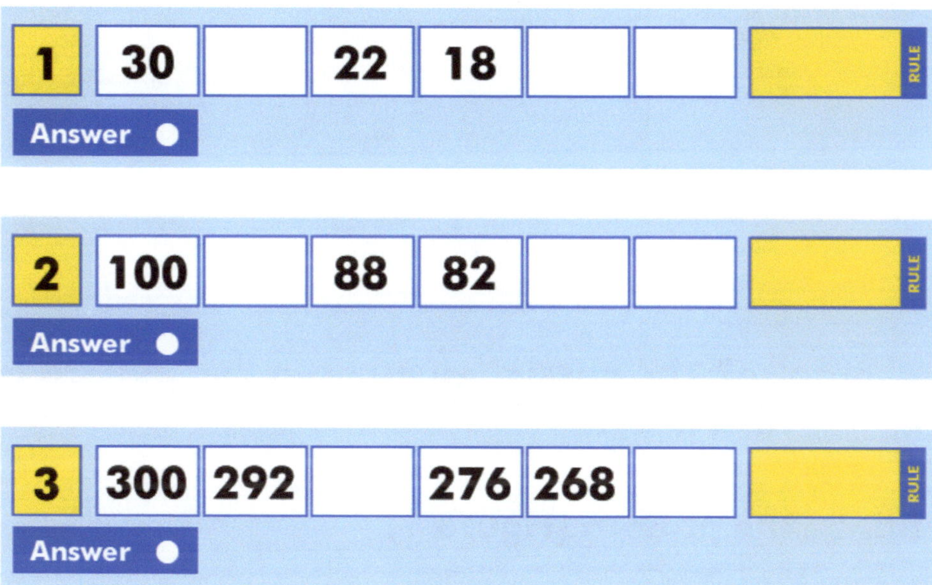

Compete the function table and find the rule.

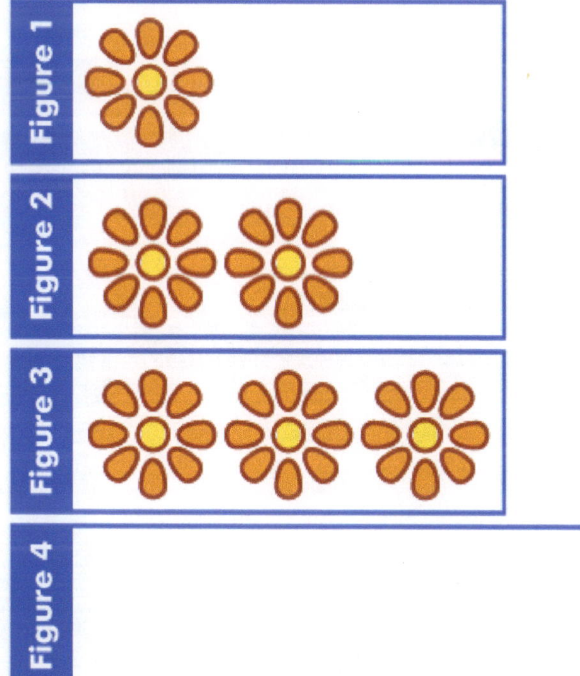

Stigmas	Petals
1	8
2	16
3	24
4	
5	
10	

Patterns and Functions Quiz

1   **True or false? The missing number in this pattern is 13.**
3, 8, ___, 18, 23, 28.

2   **Continue the pattern (Figure 1):**

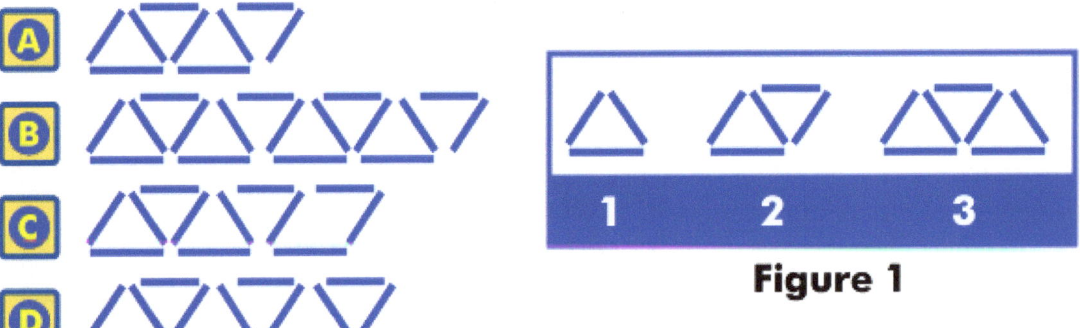

Figure 1

A

B

C

D

3   **Find the missing number in this pattern:**
___, 120, 140, 160, 180

4   **Find the missing number in this pattern:**
66, 62, ___, 54, 50

The Coordinate Plane

Key Vocabulary

coordinate plane

quadrant

origin

ordered pair

The Coordinate Plane
The numbers representing the positions on the x and y coordinates are called ordered pairs.

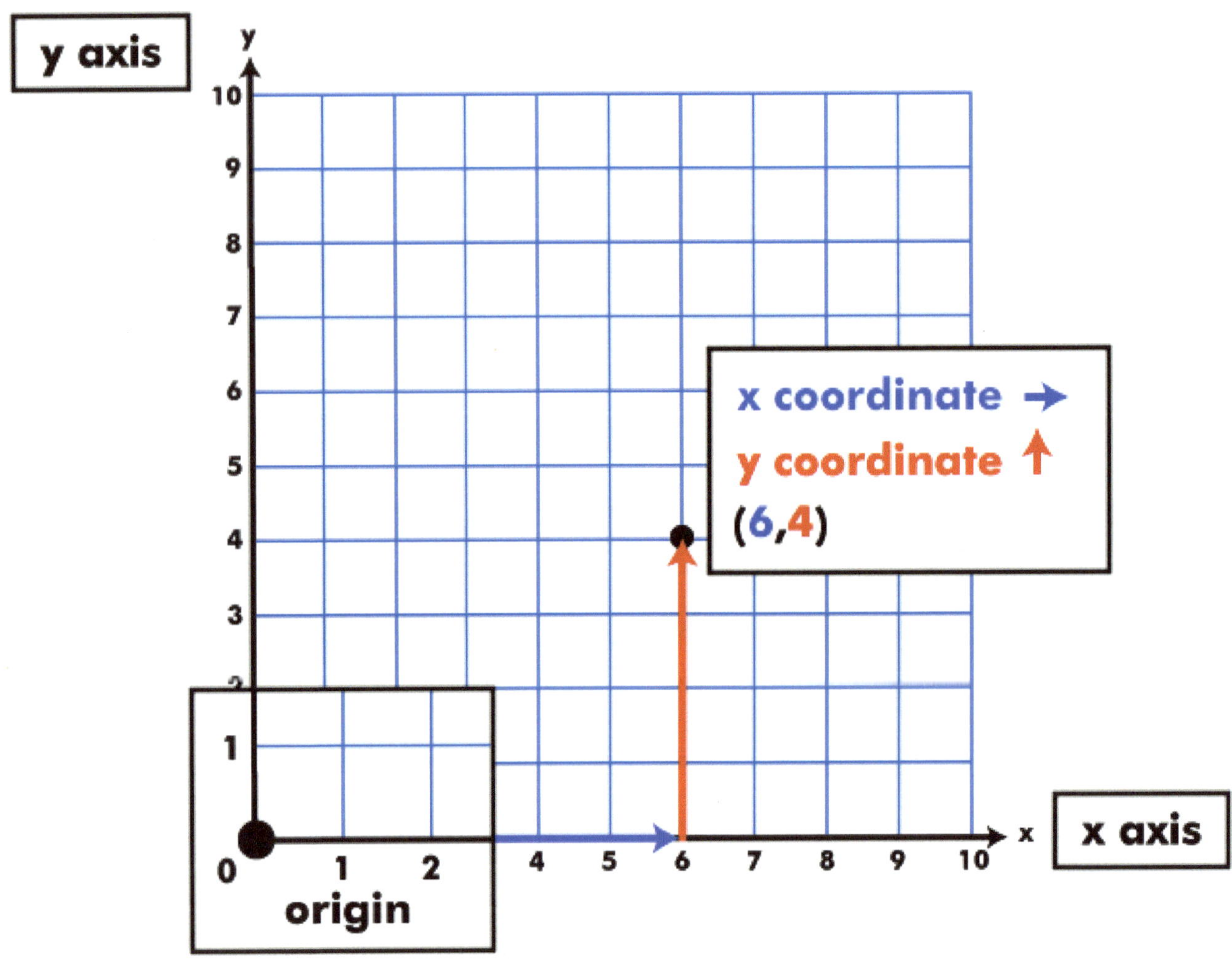

Give directions from Fenway Park (marked with a Boston B) to the New England Aquarium (marked with a fish).

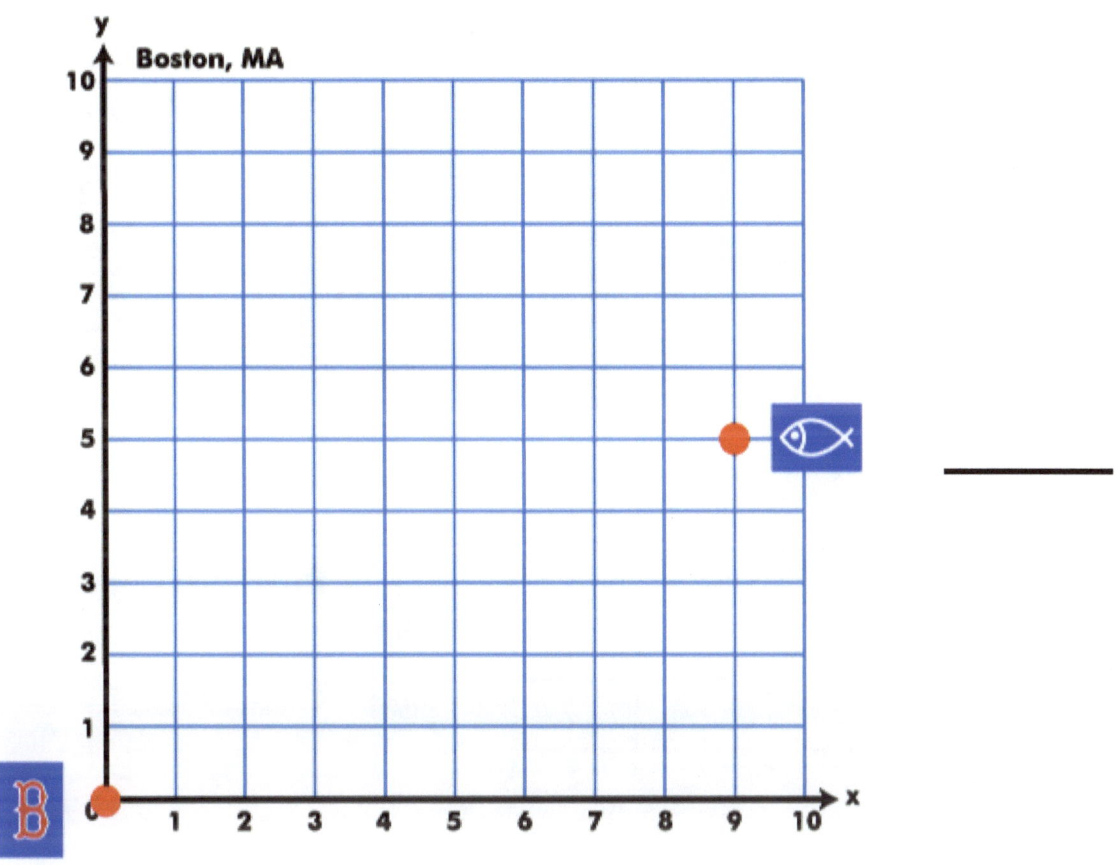

____

Hint

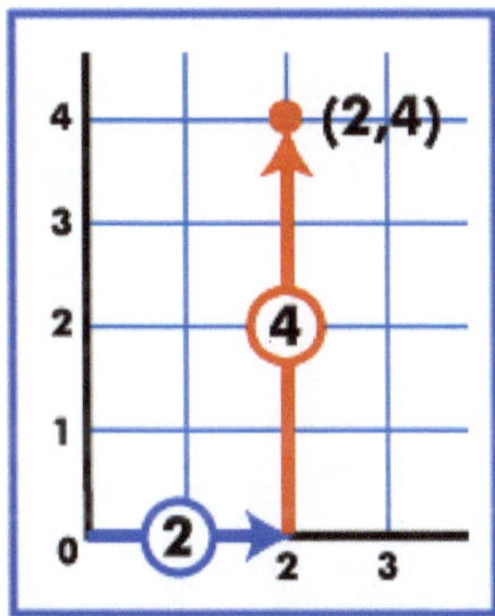

Find the ordered pairs for these other Boston landmarks.

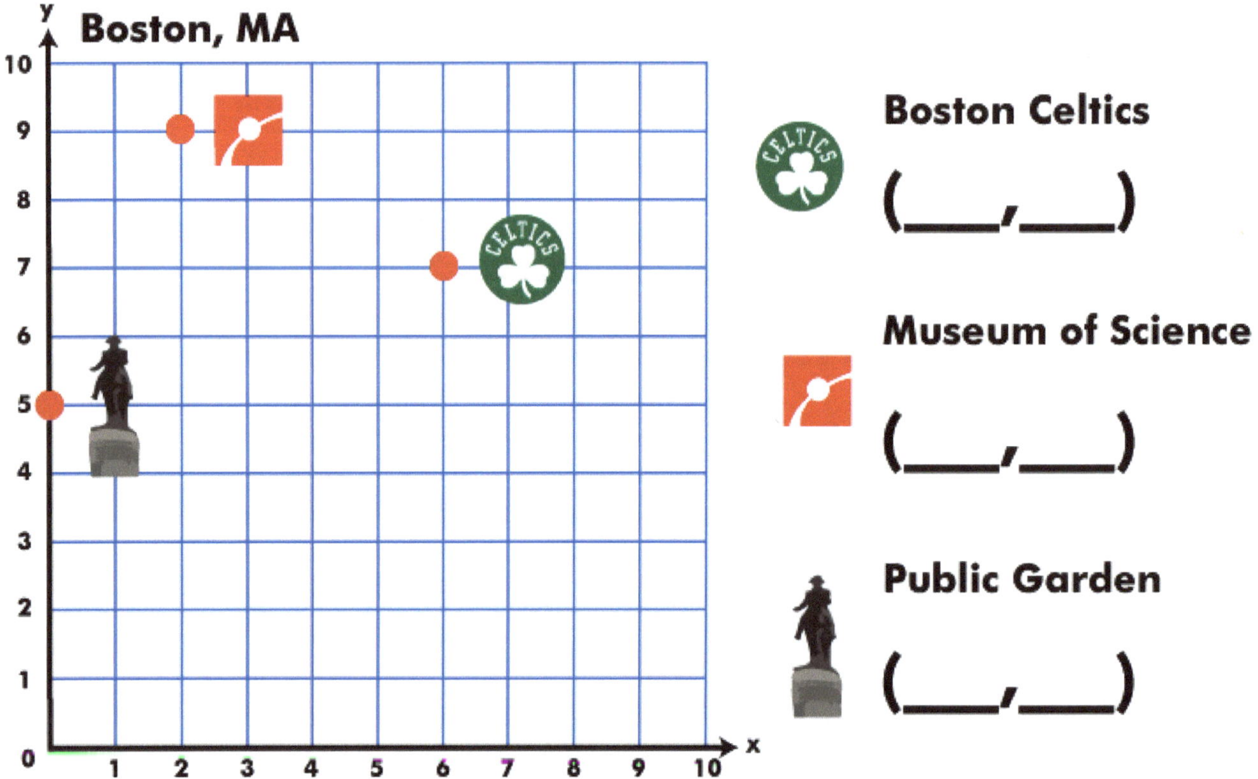

Find these Boston landmarks and plot them on the map with a blue or red dot.

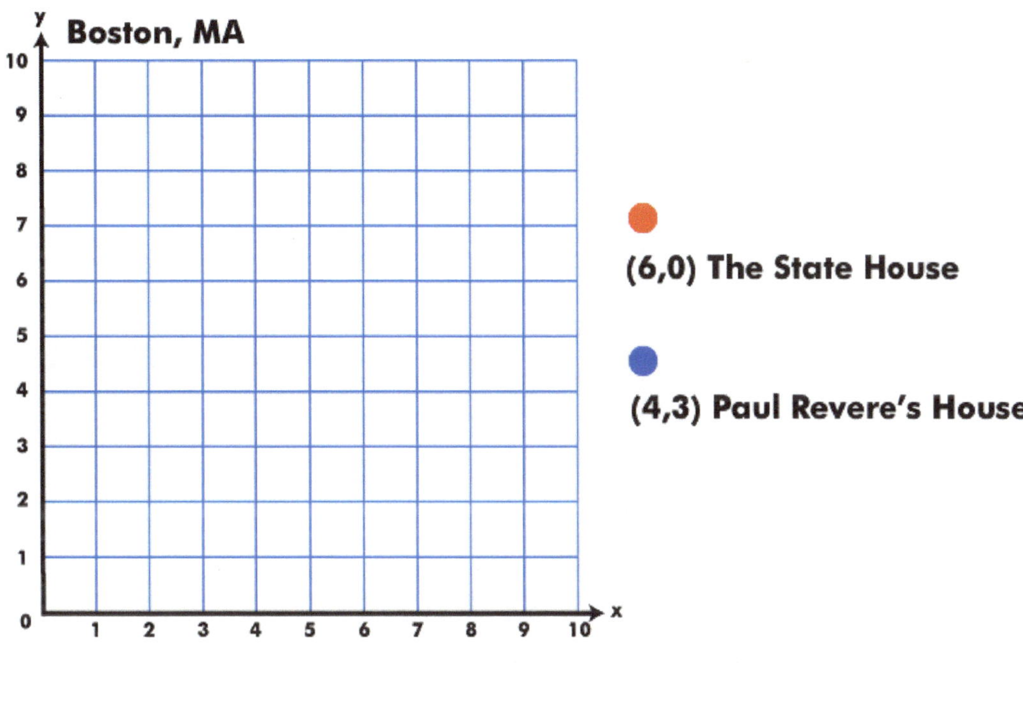

(6,0) The State House

(4,3) Paul Revere's House

Write the ordered pairs for A, B and C.

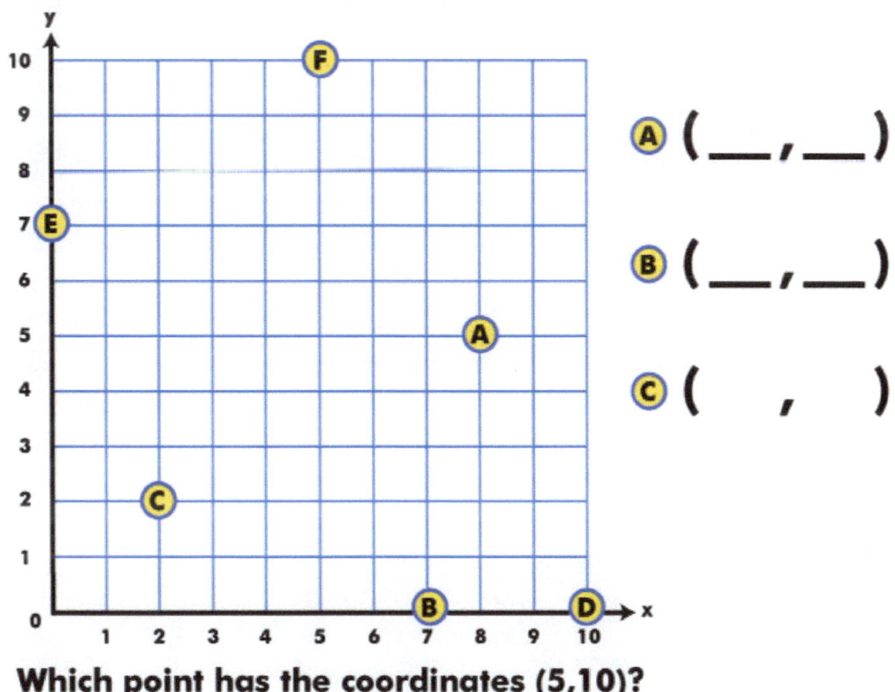

A (___ , ___)

B (___ , ___)

C (,   )

Which point has the coordinates (5,10)?

Graph these points.

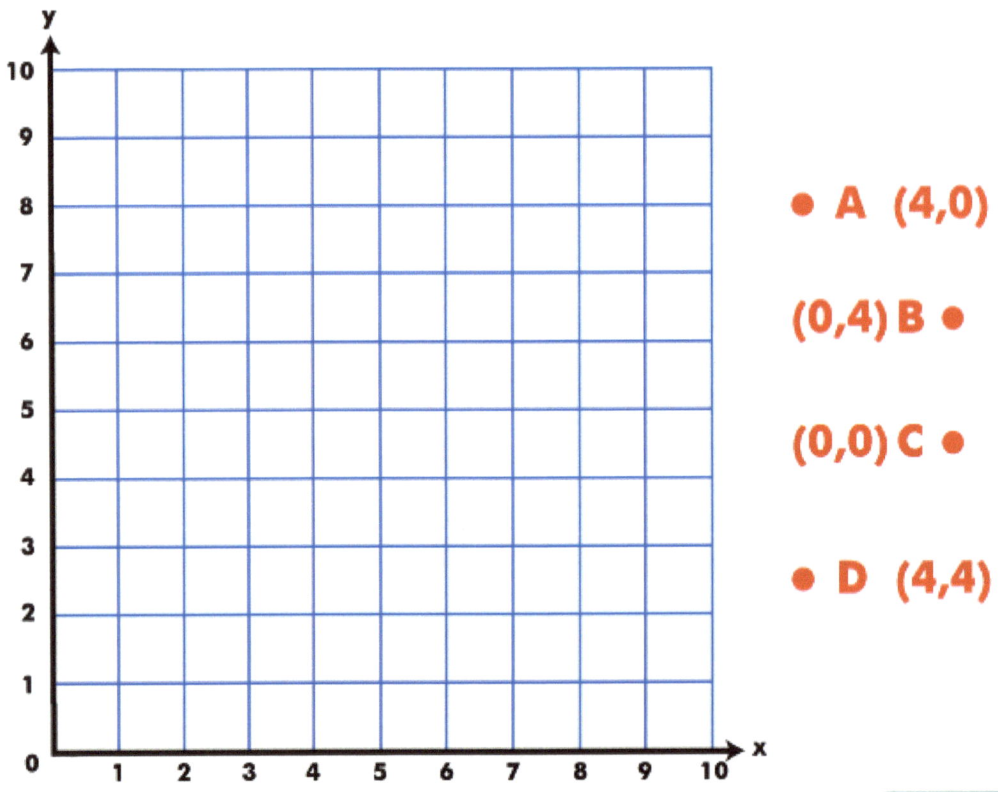

What shape will you make if you connect the points?

Name_________________________________

The Coordinate Plane Quiz

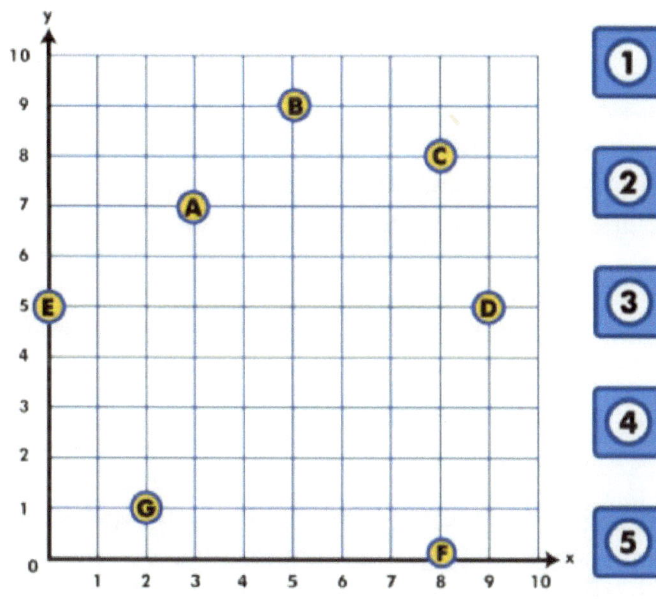

 **1** Which point has the coordinates (9,5)

 **2** Which point has the coordinates (5,9)

 **3** What is the y-coordinate of point F?

4 What is the y-coordinate of point E?

 **5** What is the x-coordinate of point E?

Newburyport, MA 01950

1-800-596-3175

OnBoard Academics employs teachers to make lessons for teachers! We create and publish a wide range of aligned lessons in math, science and ELA for use on most EdTech devices including whiteboard, tablets, computers and pdfs for printing.

All of our lessons are aligned to the common core, the Next Generation Science Standards and all state standards.

If you like our products please visit our website for information on individual lessons, teachers licenses, building licenses, district licenses and subscriptions.

Thank you for using OnBoard Academic products.